YOUR KNOWLEDGE HAS VALUE

- We will publish your bachelor's and
  master's thesis, essays and papers

- Your own eBook and book -
  sold worldwide in all relevant shops

- Earn money with each sale

Upload your text at www.GRIN.com
and publish for free

Bibliographic information published by the German National Library:

The German National Library lists this publication in the National Bibliography; detailed bibliographic data are available on the Internet at http://dnb.dnb.de .

This book is copyright material and must not be copied, reproduced, transferred, distributed, leased, licensed or publicly performed or used in any way except as specifically permitted in writing by the publishers, as allowed under the terms and conditions under which it was purchased or as strictly permitted by applicable copyright law. Any unauthorized distribution or use of this text may be a direct infringement of the author s and publisher s rights and those responsible may be liable in law accordingly.

Imprint:

Copyright © 2015 GRIN Verlag
Print and binding: Books on Demand GmbH, Norderstedt Germany
ISBN: 9783668719132

This book at GRIN:

https://www.grin.com/document/420613

Emmanuel Igweh, D. E. Peter

Acute Toxicity of Ethanolic Leaf Extract of Myrianthus Arboreus on the Liver Enzymes of Wistar Rats

GRIN Verlag

GRIN - Your knowledge has value

Since its foundation in 1998, GRIN has specialized in publishing academic texts by students, college teachers and other academics as e-book and printed book. The website www.grin.com is an ideal platform for presenting term papers, final papers, scientific essays, dissertations and specialist books.

Visit us on the internet:

http://www.grin.com/

http://www.facebook.com/grincom

http://www.twitter.com/grin_com

ACUTE TOXICITY OF ETHANOLIC LEAVES EXTRACT OF *MYRIANTHUS ARBOREUS* (P.BEAUV) ON THE LIVER ENZYMES OF WISTAR RATS

BY

IGWEH, EMMANUEL CHIDIKE
A PROJECT RESEARCH SUBMITTED TO THE DEPARTMENT OF BIOCHEMISTRY
OF UNIVERSITY OF PORT HARCOURT, RIVERS STATE, NIGERIA.

IN
PARTIAL FULFILMENT OF THE REQUIREMENT FOR THE AWARD OF BACHELOR
OF SCIENCE DEGREE (B.Sc.) IN BIOCHEMISTRY
UNIVERSITY OF PORT HARCOURT, NIGERIA.

December 2015

ABSTRACT

The present study was done to evaluate the acute (14 days) toxicity of the ethanolic leaf extract of *Myrianthus arboreus* on the liver enzymes of wistar rats. In the acute (14 days) toxicity studies, 24 rats were grouped into 1- 8groups (n=3rats/cage) and administered with 1500, 1000 and 500mg/kg body weight for 7days and 14days. The rats were sacrificed after 7days and 14day of administration and blood samples and liver organ were collected for investigations. The biochemical parameters such as the Alkaline phosphatase (ALP), Alanine transaminase (ALT) and Aspartate aminotransferase (AST) were determined and the liver histology analysed. The mean values of ALP showed significant increase (P≤0.05), the ALT showed a non-significant increase (P≥0.05) at groups 2, 3 and 4 and a significant increase (p≤0.05) at groups 6, 7 and 8. The AST showed a non-significant increase (P≥0.05) at all dosages and times except for group 2. The histological analysis showed microvesicular steatosis at groups 2 and 3 and a ballooning hepatic necrosis at group7. The phytochemical analysis of *Myrianthus arboreus* shows the presence of alkaloids, flavonoids, tannins, anthraquinones, triterpenoids, carbonhydrate, cardenolide and saponins in detectable limits but fixed oils and cyanogenic glycosides were not determined. In this investigation, we can conclude that the ethanolic leaf extract of *Myrianthus arboreus* was unsafe at all doses considered for a period of 14days. However, at a dose below 500mg for 7days could be considered safe.

LIST OF ABBREVIATIONS

AChE- Acetyl cholinesterase

ALP/ AP - Alkaline Phosphatase

ALT- Alanine aminotransferase

ANOVA- Analysis of Variance

AST- Aspartate aminotransferase

BChE/BuChE- Pseudocholinesterase

B.W – Body Weight

ChE- Choline Esterase

CYPS- Cytochrome p450 Isoenzymes

DMSO- Dimethyl Sulfoxide

EC- Enzyme Classification

ELISA- Enzyme Linked Immuno-sorbent Assay

ER- Epoxide hydrolase

FMO- Flavin monoxygenase

GGT- γ-Glutamyl Transpeptidase

GST- Gluthatione-s-transferases

IU/L - International Units/Litre

IUPAC- International Union of Pure and Applied Chemist

LD$_{50}$- Lethal Dose

LDH- Lactate Dehydrogenase

KG - Kilogram

MSDS- Material Safety and Data Sheet

M±SEM- Mean ± Standard Error of Mean

MT- Methyl transferases

NAD^{+} - Nicotinamide Adenosine Dinucleotide

NAT- n-acetyl transferases

OECD- Organization for Economic Cooperation and Development

OCT- Ornithine carbamyl transferase

OTC- Ornithine transcarbamylase

SDH- Sorbitol dehydrogenase

SGPT - Serum Glutamate-Pyruvate Transaminase

SULT- Sulphotransferases

UGT- UDP-glucuronosyl transferases

U/L-   Units/Litre

Table of contents

LIST OF TABLES

LIST OF FIGURES

LIST OF PLATES

APPENDIX

CHAPTER ONE

1.0 INTRODUCTION AND LITERATURE REVIEW

1.1 PLANT PRODUCTS AS DRUGS

Historically, natural products have remained the major and potent source of new drugs. The use of natural products with therapeutic properties is as ancient as human civilisation and, for a long time, mineral, plant and animal products were the main sources of drugs (De Pasquale, 1984). Natural products which included herbs, animals and minerals serve as the lead compounds for the development of new medicines and also for the treatment and prevention of various human ailments.(Prasanth *et al., 2014*).

Plant products are classified into; Primary metabolites and secondary metabolite.

i. Primary metabolites are chemical compounds (metabolites) that are essential for the basic metabolism (anabolism and catabolism) of the plants, which results to assimilation, respiration, transport, and differentiation. They incudes starch, amino acids, lipids, minerals and vitamins.

ii. Secondary metabolites are chemical compounds (metabolites) that are non-essential for the basic metabolism of the plants. These products accounts for the colour, flavours and smells and are source of fine chemicals such as drugs, insecticides, dyes, fragrances and the phyto-medicines found in medicinal plants. They include alkaloids, glycosides, terpenes/isoprenoids, amines and phenolics.

Secondary metabolites (which are the potent forms of the natural products) and their derivatives, serve as the structural template for discovery and development of new drugs. The discovery and development of drug remains a challenging scientific task, which is the transition from a screening hit to a drug candidate, requires expertise and experience (Mouhssen, 2013). Although products derived from natural sources may not necessarily represent active ingredients in their final form, the majority of all drugs in the market have their origin in nature. (Chin *et al.*,2006, Newman and Cragg 2012).

The growing number of herbal drug users around the globe and lack of scientific data on the safety profile of herbal products make it necessary to conduct toxicity study of herbal products. (Saad, 2006).

1.2 TOXICITY TESTS

Toxicity tests are most widely used to examine specific adverse events or specific endpoints such as cancer, cardio-toxicity and skin/eye irritation.(Prasanth *et al.*, 2014).

Acute, sub-acute and chronic toxicity tests are routine toxicity tests carried out by the pharmaceutical companies in the development of new medicines. Acute, Sub-acute and chronic toxicity tests are carried out to evaluate the toxic nature of a compound.

Acute toxicity testing involves the determination of lethal dose, the dose that kills 50% of the tested group of animals, whereas sub-acute and chronic toxicity testing involves the determination of long term effects of the test compound upon repeated administration (Prasanth *et al.*, 2014).

Acute toxicity describes the adverse effects of a substance that result either from a single exposure (The MSDS Hyper-Glossary, 2006) or from multiple exposures in a short space of time (usually less than 24 hours).

To be described as *acute* toxicity, the adverse effects should occur within 14 days of the administration of the substance. (IUPAC, 1997).

The acute toxicity test in which a single dose is used in each animal on one occasion only for the determination of gross behaviour and LD_{50} or median lethal dose. (Bhardwaj *et al.*, 2012).

Sub-acute toxicity tests are employed to determine toxicity likely to arise from repeated exposures or daily dose of several weeks to several months (usually 4 weeks to 6 months).(Bhardwaj *et al.*, 2012).

Chronic toxicity tests determine toxicity from exposure for a substantial portion of a subject's life, (usually 12 to 24 months). (Bhardwaj *et al.*, 2012).

Acute toxicity testing involves the determination of lethal dose (LD_{50}), the dose that kills 50% of the tested group of animals. *(Prasanth et al., 2014).*

Standardized tests or routes of administration of the test substance are available for oral, intravenous, intra-peritoneal, subcutaneous and inhalation exposures.

1.3   THE USE OF EXPERIMENTAL ANIMAL

It is widely considered unethical to use humans as test subjects for acute (or chronic) toxicity research. Otherwise, most acute toxicity data comes from animal testing or, more recently, *in vitro* testing methods and inference from data on similar substances. (Walum, 1998).

Fig.1.1 A Wistar Rat

The Wistar rats are currently one of the most popular rat strains used for laboratory research. It is characterized by its wide head, long ears, and having a tail length that is always less than its body length. Wistar rats are an outbred strain of albino rats belonging to the species *Rattus norvegicus*. This strain was developed at the Wistar Institute in 1906 for use in biological and medical research, and is notably the first rat strain developed to serve as model organism at a time when laboratories primarily used Mus musculus. (Iliuță, 2011).

Again, owing to the fact that the albino rats and other experimental animals have great similarities with human, they are used for experimental purposes. Thus, experimental results generated from them can be extrapolative (applicable) to humans while considering the differences in their organismal structure.

1.4   THE LIVER

1.4.1   STRUCTURE AND FUNCTIONS

The human liver is a dark red-brown organ with a soft, spongy texture. (Encarta, 2009). The liver has four lobes of unequal size and shape, weighing between 1.44- 1.6kg (Crook, 2009).

It is the largest internal organ in the human body, and is located in the right upper quadrant of the abdominal cavity, resting just below the diaphragm.

The liver is connected to two large blood vessels which accounts for the blood flow in the liver;

i. The Hepatic artery, which deliver oxygen-rich blood from the heart supplying about 25 per cent of the liver's blood.

ii. The Hepatic Portal vein, which carries blood to the liver that has travelled from the digestive tract, where it collects nutrients as food is digested. These nutrients are delivered to the liver for further processing or storage. This vein is the source of 75 per cent of the liver's blood supply. (Encarta, 2009).

The liver plays an astonishing array of vital functions in the maintenance, performance and regulating homeostasis of the body. It is involved with almost all the biochemical pathways to growth, fight against disease, nutrient supply, energy provision and reproduction (Sharma et al., 1991). The liver, which is part of the digestive system, performs more than 500 different functions, all of which are essential to life. (Encarta,2009).

The major functions of the liver are carbohydrate, protein and fat metabolism, detoxification, secretion of bile, decomposition of red blood cells, plasma protein synthesis, hormone production, cholesterol synthesis and storage of vitamins. Thus, it is now clear to say that maintaining a healthy life, the liver is a crucial factor for the overall health and wellbeing of an individual.

The liver is unique among the body's vital organs in that it can regenerate, or grow back, cells that have been destroyed by some short-term injury or disease. However, if the liver is damaged repeatedly over a long period, it may undergo irreversible changes that permanently interfere with function. (Encarta, 2009)

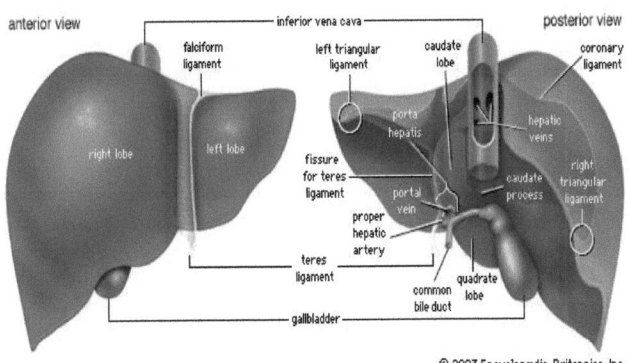

© 2003 Encyclopædia Britannica, Inc.

Fig. 1.2 Diagram of the liver (Encyclopaedia Britannica, 2003)

The liver is composed of cell called hepatocytes. There are two major types of cells that populates the liver globes;

i. The karat parenchymal cell, which covers 80% of the liver.

ii. The non-parenchymal cell which includes, sinusoidal hepatic endothelial cells, kupffer cells and hepatic stellate cells, covers 40% of the liver (Kmie,2001).

1.4.2 HEPATOTOXICITY

The liver is the most common site of damage in laboratory animals administered drugs and other chemicals. There are many reasons including the fact that the liver is the first major organ to be exposed to ingested chemicals due to its portal blood supply. Although chemicals are delivered to the liver to be metabolized and excreted, this can frequently lead to activation and liver injury. (Andy, 2001).

Hepatotoxicity implies chemical-driven liver damage. Certain medicinal agents, when taken in overdoses and sometimes even when introduced within therapeutic ranges, may injure the organ. Chemicals that cause liver injury are called hepatotoxins. Other chemical agents, such as those used in laboratories and industries, natural chemicals (e.g. microcystins) and herbal remedies can also induce hepatotoxicity. (Aashish *et al.*, 2011).

Several mechanism are responsible for either inducing hepatic injury or worsening the damage process. Herbal remedies, synthetic drugs and some chemicals act by damaging the mitochondria of the hepatocytes (Pak *et al.*, 2004). Its dysfunction releases excessive amount of oxidants, which in turn injure the hepatic cells. Activation of some enzyme in the cytochrome P450 systems also leads to oxidative (Lynch *et al.*, 2007). Injury to the hepatocyte and bile duct cells leads to accumulation of bile acid inside the liver this promotes further liver damages.

1.4.3 DETOXIFICATION FUNCTION OF THE LIVER

The various functions of the liver are carried out by the liver cells or hepatocytes (Benjamin and Sherman, 2008). Amongst the many metabolic functions of the liver, the most important and probably essential function is in the detoxification of xenobiotic or foreign chemicals. Today, most of xenobiotic to which humans are exposed to come from sources that includes; environmental pollution, food additives, cosmetic products, agrochemicals, processed foods and drugs. The detoxification system of liver involves two phases viz; Phase I and Phase II.

The Phase I Process involves exposure of the xenobiotic to functional groups (amino, carboxylic, hydroxyl, thiol, etc.) which brings about reduction and oxidation of the xenobiotic. The process is catalysed by Cytochrome P450 isoenzymes (cyps), flavin monoxygenases (FMO) and Epoxide hydrolases (ER).

The Phase II process involve conjugation of the phase one products and consequently making them hydrophilic (water soluble). The conjugation enzyme includes; gluthation-s-transferase (GST), UDP-glucuronosyl transferase (UGT), Sulphotranferases (SULT), n-acetyl transferases (NAT) and methyl transferases (MT).

1.4.4 MARKER MOLECULES AND LIVER TOXICITY

Every cellular organisational level (organelle, cell, tissue, organ and system) have molecules that are specific to it. These molecules such as enzymes, hormones, proteins are called 'Marker Molecules'. The distribution of marker enzymes in the cell reflects the compartmentation of the processes they catalyze. (koolman *et al.*, 2005).

The liver has marker molecules such as enzymes, hormones and proteins. Marker enzymes of the liver are determinant factor in assessing the levels of liver injury (liver injury test).

Liver toxicity can be classified into Cholestasic Injury, Cytotoxic Injury and Disturbances of hepatic function/clearance.

Similarly, Evaluation of liver toxicity in vivo is carried out by; Serum enzyme tests, Hepatic excretory tests, Alterations in chemical constituents of the liver and Histological analysis of liver injury.

The marker enzymes involved in the liver injury test include; Alkaline Phosphatase [AP, ALP], γ-Glutamyl Transpeptidase [GGT], Alanine aminotransferase [ALT], Aspartate aminotransferase [AST], Lactate Dehydrogenase [LDH], Ornithine carbamyl transferase [OCT], Sorbitol dehydrogenase [SDH], and Choline Esterase [ChE]. (Enzyme markers of Enzyme markers of Toxicity, 2015).

i.   **Alkaline phosphatase (ALP)** (EC 3.1.3.1) is a hydrolase enzyme responsible for removing phosphate groups from many types of molecules, including nucleotides, proteins, and alkaloids. The process of removing the phosphate group is called dephosphorylation. As the name suggests, alkaline phosphatases are most effective in an alkaline environment. In humans, alkaline phosphatase is present in all tissues throughout the entire body, but is particularly concentrated in *liver,* bile duct, kidney, bone, intestinal mucosa and the placenta.

ii. **Gamma-glutamyl transpeptidase (GGT)** is an enzyme that catalyses the reaction between a peptide and an amino acid. High concentrations are found in the *liver* and kidney. GGT is measured in combination with other tests. ALP is increased in hepatobiliary disease and bone disease; GGT is elevated in hepatobiliary disease, but not in bone disease.

iii. **Alanine transaminase (ALT)** is a transaminase enzyme (EC 2.6.1.2). It is also called Alanine aminotransferase (ALAT) and was formerly called serum glutamate-pyruvate transaminase (SGPT) or serum glutamic-pyruvic transaminase (SGPT). ALT is found in plasma and in various body tissues, but is most common in the *liver.* It catalyses the two parts of the Alanine Cycle. ALT is commonly measured clinically as a part of a diagnostic evaluation of hepatocellular injury, to determine liver health. When used in diagnostics, it is usually measured in international units/litre (IU/L). Significantly elevated levels of ALT often suggest the existence of other medical problems such as viral hepatitis, diabetes, congestive heart failure, liver damage, bile duct problems, infectious mononucleosis, or myopathy, so ALT is commonly used as a way of screening for liver problems. Haemolysis has a negligible effect on ALT activity

iv. **Aspartate transaminase (AST)** or **Aspartate aminotransferase** (EC 2.6.1.1). AST catalyses the reversible transfer of an α-amino group between aspartate and glutamate and, as such, is an important enzyme in amino acid metabolism. AST is found in the *liver,* heart, skeletal muscle, kidneys, brain, and red blood cells. Serum AST level, serum ALT (alanine transaminase) level, and their ratio (AST/ALT ratio) are commonly measured clinically as biomarkers for liver health. The tests are part of blood panels. AST is commonly measured clinically as a part of diagnostic liver function tests, to determine liver health. However, it is important to keep in mind that the source of AST (and, to a lesser extent, ALT) in blood tests may reflect pathology in organs other than the liver. In fact, when the AST is higher than ALT, a muscle sourcing of these enzymes should be considered. For example, muscle inflammation due to dermatomyositis may cause AST>ALT. This is a good reminder that AST and ALT are not good measures of liver function because they do not reliably reflect the synthetic ability of the liver and they may come from tissues other than liver (such as muscle).

8

v. **Lactate dehydrogenase (LDH or LD)** is an enzyme found in nearly all living cells (animals, plants, and prokaryotes). LDH catalyses the conversion of pyruvate to lactate and back, as it converts NADH to NAD^+ and back. A dehydrogenase is an enzyme that transfers a hydride from one molecule to another. LDH has been of medical significance because it is found extensively in body tissues, such as blood cells, *liver* and heart muscle. Because it is released during tissue damage, it is a marker of common injuries and disease such as heart and liver failure.

vi. **Sorbitol dehydrogenase** (SDH) is a cytosolic enzyme. Sorbitol dehydrogenase is an enzyme in carbohydrate metabolism converting sorbitol, the sugar alcohol form of glucose, into fructose. Organs that use it most frequently include the *liver* and seminal vesicle. It is a Sensitive enzyme marker for liver necrosis but shall be combined with measurements of ALT or other enzymes. It is elevated after acute obstruction of bile flow.

vii. **Cholinesterase** is a family of enzymes that catalyse the hydrolysis of the neurotransmitter acetylcholine into choline and acetic acid, a reaction necessary to allow a cholinergic neuron to return to its resting state after activation. There are two forms of cholinesterase;

- **Acetyl cholinesterase** (EC 3.1.1.7) (AChE) is found primarily in the blood on red blood cell membranes, in neuromuscular junctions, and in neural synapses. Acetyl cholinesterase exists in multiple molecular forms.
- **Pseudocholinesterase** (EC 3.1.1.8) (BChE or BuChE), also known as *plasma cholinesterase, butyrylcholinesterase,* or (most formally) *acylcholine acylhydrolase,* is produced in the *liver* and found primarily in plasma. Pseudocholinesterase levels may be reduced in patients with advanced liver disease

viii. **Ornithine transcarbamylase** (OTC) (also called **Ornithine carbamoyl transferase**) is an enzyme that catalyzes the reaction between carbamoyl phosphate (CP) and Ornithine to form Citrulline and Phosphate. In plants and microbes, OTC is involved in Arginine (Arg) biosynthesis, whereas in mammals it is located in the mitochondria and is part of the urea cycle. It is found in liver (>97%) and small intestine (<2%) and its activity increases in both acute and chronic liver disease.

Notably, when elevated ALT levels are found in the blood, the possible underlying causes can be further narrowed down by measuring other enzymes. For example, elevated ALT levels due to hepatocyte damage can be distinguished from bile duct problems by measuring alkaline phosphatase. In addition, myopathy-related elevations in ALT should be suspected when the aspartate transaminase (AST) is greater than ALT; the possibility of muscle disease causing elevations in liver tests can be further explored by measuring muscle enzymes, including Creatine Kinase. Many drugs may elevate ALT levels, including Zileuton, omega-3-acid ethyl esters (Lova za), anti-inflammatory drugs, antibiotics, cholesterol medications, some antipsychotics such as risperidone, and anticonvulsants. Paracetamol may also elevate ALT levels. (Wikipedia, 2015).

In fact, abnormal level (especially elevation) of these liver marker enzymes in the serum interprets liver injury.

1.5 *MYRIANTHUS ARBOREUS,* ORIGIN, CLASSIFICATION AND MEDICAL USES

Myrianthus arboreus is a common tree in the forest area of West and Central Africa, occurring in rain forest, semi-deciduous forest and swamp forest. (Okafor, 2004). It belongs to the family Cecropiaceae and it is a Dioecious shrub or tree up to (14–20) m tall. (*P.beauv.*).The indigenous wild plant is used for food and medicine.

Fig.1.3a.The Tree of *Myrianthus arboreus*
Source: Camera Photograph.

Fig.1.3b. The leaves of *Myrianthus arboreus*
Source: Camera Photograph.

The sweet or acidulous pulp around the seed is edible and the young leaves are eaten as vegetable. The leaves, bark, leafy shoots and roots are used for medicine. In Rumuji, Emohua local Government area of Rivers state of Nigeria, the leaves are cooked, the steam from the pot is inhaled, and afterwards the hot aqueous extract is used for bathing and this offers a wide range pain (analgesic) relief. Ethno-botanically, it is called 'Uzere' in Ikwerre- Rumuji language in Rivers state of Nigeria.

A bark decoction is drunk to treat malaria, fever and cough. Extracts of the leaves or leafy shoots of *Myrianthus arboreus* are used in Sierra Leone, Nigeria and the Mount Cameroon area in preparations to treat dysentery, diarrhoea and vomiting.

In the Igala area of Nigeria, the leaves are an ingredient of a febrifuge given to young children. (Okafor, 2004).

Within the continents of African such as Nigeria and Congo, chopped leaves are eaten raw with salt for heart troubles, pregnancy complications, dysmenorrhea, incipient hernia and a plaster made of beaten leaf applied to boils. Sap from the leaves is applied topically for toothache, to the chest for bronchitis or as throat pain for sore throat.

In Tanzania, an infusion of the leaves is taken to improve lactation in women. (Okafor, 2004).

1.6 TAXONOMICAL CLASSIFICATION OF *MYRIANTHUS ARBOREUS*

Kingdom: Plantae
Phylum: Tracheophyta
Class: Magnoliopsida
Order: Urticales
Family: Cecropiaceae
Genus: Myrianthus
Species: Arboreus
Scientific name: *Myrianthus arboreus*

Data from source **Arctos Plants** "http://arctos.database.museum/includes/style.min.css"

1.7 COMPOSITIONS AND PHYTOCHEMISTRY OF *MYRIANTHUS ARBOREUS*

The composition of fresh fruit pulp of *Myrianthus* sp. per 100 g edible portion is: water 85.5 g, energy 205 kJ (49 kcal), protein 1.9 g, carbohydrate 11.8 g, Ca 44 mg, P 70 mg, Fe 1.1 mg.
The composition of dried seeds per 100 g is: water 13.5 g, energy 1972 kJ (471 kcal), protein 23.6 g, fat 33.4 g, carbohydrate 27.0 g, fibre 3.5 g, Ca 132 mg, P 371 mg, Fe 6.6 mg (Leung *et al.*, 1986).
The oil consists almost exclusively of linoleic acid (93%). The protein is rich in the amino acid cysteine, which is important in a region where chronic deficiency of sulphur-bearing amino acids occurs. Several pentacyclic triterpenoids have been isolated from the wood and the roots. Euscaphic acid, myrianthic acid, tormentic acid, ursolic acid and a derivative of ursenoic acid have been isolated from stems. Myrianthinic acid was isolated from the bark. The wood also contains myrianthiphyllin, a lignan cinnamate. Bark extracts of *Myrianthus arboreus* showed antiplasmodial, antimycobacterial and antitrypanosomal. (Okafor, 2004). The methanol leaf extract of *Myrianthus arboreus* as well as the aqueous and ethanol exhibited antimicrobial and antioxidant activities.(Agyare *et al.*,2014).

The phytochemical screening done by (Oyeyemi *et al.*, 2014) revealed that the leaves of *Myrianthus arboreus* has;
  i.   Flavonoids(45.62 ± 0.07)
  ii. Alkaloids(40.56 ± 0.05)
  iii. Saponins(0.00 ± 0.00)
  iv. Tannins(6.88 ± 0.02)
  v.   Phenols(7.02 ± 0.02)
  vi. Glycosides(4.60 ± 0.01)

Similarly, the proximate analysis done by (Oyeyemi *et al.*,2014) showed that the leaves of *Myrianthus arboreus* has;

  i.   Protein ($4.20 \pm 0.00\%$)
  ii. Crude fat/Lipid ($6.01 \pm 0.10\%$)
  iii. Crude fibre ($4.71 \pm 0.00\%$)
  iv. Moisture content ($35.20 \pm 0.84\%$)
  v.   Ash content ($3.52 \pm 0.18\%$)
  vi. Carbohydrate ($7.20 \pm 0.14\%$)
  vii. Dry matter ($8.10 \pm 0.14\%$)

viii.   Calcium (90.08 ± 0.60 mg/100g)
ix.   Magnesium(2.68 ± 0.17 mg/100g)
x.   Sodium (17.95 ± 0.07 mg/100g)
xi.   Potassium (70.62±0.39 mg/100g)
xii.   Iron (7.44±0. 15 mg/100g).

Despite the various uses over long time, no toxicological data is available regarding the safety of repeated exposure to *Myrianthus arboreus*. Considering the numerous nutritional and health benefits, it is therefore necessary to conduct a toxicological study on *Myrianthus arboreus* and ascertain data regarding its safe use.

1.8   AIM AND OBJECTIVES OF THE STUDY

AIM:
Aim is centred on the acute toxicity of ethanolic leaf extracts of *Myrianthus arboreus* (P.Beauv) on the liver enzymes of wistar rats

OBJECTIVES:
The objectives of this study are to;
   i.   Evaluate the LD_{50} of *Myrianthus arboreus* on wistar rats.
   ii.   Qualitatively determine the phytochemicals of the *Myrianthus arboreus*.
   iii.   Determine the activities of Alkaline Phosphatase (ALP), Alanine Transaminase (ALT) and Aspartate Aminotransaminase (AST) in the liver of wistar rats.

CHAPTER TWO

2.0   MATERIALS AND METHODS

2.1   MATERIALS

2.1.1   LABORATORY APPARATUS

Aluminium foil, Beakers, Cotton wool, Crucibles, Desiccator, Dissecting kits, Face mask, Filter paper(Whatman), Funnel, Gavage syringe, Glass rod, Hand gloves(plastic and elastic), Laboratory coat, Marceration Jar (5000ml), Microtome, Reagent bottles, Spatula, Sterile specimen tubes and Syringes.

2.1.2   LABORATORY EQUIPMENT

Analytic weighing Balance (HCK LN 0708)
Centrifuge (Universal 320 laboratory century Hettich Zentrifugen)
Digital water bath (TT-6 Techmel & Techmel USA)
Digital weighing balance (Satorius TE1535, CANADA)
Electronic Blender
Heating incubator (DHP-9053A)
Microscope (Photomax (LB) Premier universal microscope 1966 OLYMPUS TOKYO)
Microtome (American Optical Microtome 820)
Rotary evaporator (RE-52A: ENGLAND LAB SERVICES)

Spectrophotometer (Surgispec 5M-23D: SURGIFEILD MEDICAL, ENGLAND).

2.1.3 CHEMICAL AND REAGENTS

Absolute Ethanol (99.7%) (JHD), Chloroform, Concentrated Dimethyl Sulfoxide (DMSO), Einosine, Formalin (10%), Heparin blood tube, Paraffin, Tap water, Xylene.

2.1.4 SOURCE OF ANIMAL

Thirty five (35) Albino Rats animals were used for the study. The rats were purchased from the Animal house of Pharmacology Department of the Institution where the experiment was also carried out.

Animal feed and tap water within the animal house were used to feed the animals.

The animals were restrained in a Plastic cage matted with sawdust.

2.2 METHODS

2.2.1 COLLECTION AND IDENTIFICATION OF PLANT MATERIALS

The fresh leaves of *Myrianthus arboreus* were collected from Rumuji-odegu in Emohua Local Government Area, Rivers State, Nigeria in August 2015. Dr. Chimezie Ekeke, a botanist at the Department of Plant Science and Biotechnology, University of Port Harcourt, Rivers State, Nigeria identified and taxonomically authenticated the plant sample. A voucher-specimen of the collected sample with voucher-specimen number UPH/V/1,224 has been deposited in the herbarium of the institution for subsequent references. See Appendix 3 for herbarium number.

2.2.2 PREPARATION OF EXTRACT

The leaves of *Myrianthus arboreus* were rinsed in clean water, shade-dried for two (2) weeks, milled using electronic blender into coarse powder, weighed and extracted with 99.7% ethanol (JHD) by cold maceration method for 72hours with intermittent shaking.

The extract was prepared in a ratio of 1: 4 i.e. 1gram of the powder to 4ml of ethanol. 200g of the fine powder was soaked in 800ml of the solvent (absolute Ethanol) inside a 5000ml maceration jar at room temperature (28 ± 2^0C), shook every 30mins for 6hours and allowed to stand for about 72hours. Using a sterile muslin cloth followed by a funnel-filter paper filtration, the solution was filtered into a sterile container. The marc was re-soaked twice in ethanol (with ethanol covering the marc) for 24hours in each case. The entire solution was concentrated and left to evaporate to dryness using a Rotary evaporator (45°C) at the Pharmacognosy and Phytotherapy laboratory, Faculty of Pharmaceutical Sciences of the institution. The extract was purified in a crucible on a water bath to dry (45°C) for 24hours. The dark green to black gel-like crude extract was preserved in an airtight sterile container and refrigerated until required. (A modified method described by Agwa, O.K. *et al*).

2.2.3 PHYTOCHEMICAL SCREENING

Phytochemical screening of *Myrianthus arboreus* leaf extract was done in the Pharmacognosy and Phytotherapy Department, Faculty of Pharmaceutical Sciences of the institution. This was done to identify the secondary metabolite using standard procedures.

The following parameters were determined in the crude extract of *Myrianthus arboreus;*

   i.   **Alkaloids:** Drangedorff's test, Mayer's test, Hager's test.

   ii.   **Flavonoids:** Shinodu's test, Leadacetaete test, AICB test, NaOH.

   iii.   **Tannins:** FeCB test, Phlobatannins, Gelatin test, Albumin test.

   iv.   **Anthraquinone (BontragersTest):** Free Anthraquinone, Combined Anthraquinone.

v.   **Triterpenoids/Steroids:** Liebermann-Buchard test, Salwoski test.

vi.   **Fixed oils**

vii.   **Carbonhydrates:** Molisch test. Fehlings test.

viii.   **Cardenolide:** Keller Killani test Kedde test

ix.   **Cyanogenic glycosides**

x.   **Saponins:** Frothing test, Haemolysis test, Emulsion test.

2.2.4   SELECTION AND SORTING (INTO SIZE AND SEX) OF ANIMALS

The Wistar rats (weighing between 90.10g and 270.20g) of the both sexes were selected and sorted into groups (with weight difference not more than 5g) for Lethal Dose (LD$_{50}$) and Acute toxicity studies. Two weeks prior to the period of the study, the selected animals were acclimatized. During this period, the animals were kept under favourable laboratory conditions and had access to food and water at the Animal house of Pharmacology Department of the Institution. The animal house attendant communicated the ethical protocol of handling the animals.

2.2.5   RECONSTITUTION OF EXTRACT

The extract was reconstituted with concentrated Dimethyl sulfoxide (DMSO). This was owing to the fact that DMSO, (a polar aprotic solvent) dissolves both polar and nonpolar compounds and is miscible in a wide range of organic solvents as well as water. It acts as a good carrier for other substances or drugs and it potentiates their effect. In addition, drugs are able to pass through the blood-brain barrier, which is usually impenetrable. (Segura, 2011)

2.2.6   LETHAL DOSE (LD$_{50}$) STUDY OF THE ETHANOLIC LEAF EXTRACTS OF *MYRIANTHUS ARBOREUS*

A lethal dose (LD50) study was carried out using an Up and down method. In which case, Nine (9) animals weighing between 163g- 263g were grouped into three having three animals in each group. The groups were designated A, B and C.

Group A administered with 5000mg/Kg Body weight.

Group B administered with 3000mg/kg Body weight.

Group C administered with 1000mg/kg Body weight.

Before the administration, the animals were made to fast for 2hours. In addition, the single dose of the plant extract was administered orally via a gavage syringe.

In each case above, 1ml of DMSO was used as a carrier.

After administration of the extract, the animals were observe for general body changes, symptoms of toxicity and mortality for the first four (critical) hours, then over a period of 24 hours, thereafter daily for 7days.

No death was observed; safe doses of 500, 1000 & 1500 mg/Kg Body weight were selected for the research. This was done with the procedures described by (OECD, 2008).

2.2.7   EXPERIMENTAL DESIGN

ACUTE TOXICITY (14 DAYS) STUDY:

In this study, the test animals were grouped into Eight (8) groups, having three (3) animals each. The groups were designated; Groups I – VIII. The dosages of administration were as follows;

Table 2.1 Experimental Design

Groups	Dosage of Administration per Body weight	Number of Rats per cage	Days of Administration
Group I	Control (DMSO)	3	7days
Group II	1500mg/Kg Body weight	3	7days
Group III	1000mg/Kg Body weight	3	7days
Group IV	500mg/Kg Body weight	3	7days
Group V	Control (DMSO)	3	14days
Group VI	1500mg/Kg Body weight	3	14days
Group VII	1000mg/Kg Body weight	3	14days
Group VIII	500mg/Kg Body weight	3	14days

Table 2.2 Administration of *Myrianthus arboreus* (MA) Extracts

Groups	Average weight of Animals (grams)	Calculated Quantity of MA extract Administered (grams)	Days of Administration
Group I	121.3g	1ml of the carrier (DMSO)	7days
Group II	123.3g	0.1849g of the extract dissolved in 1ml of the carrier (DMSO)	7days
Group III	120.5g	0.1205g of the extract dissolved in 1ml of the carrier (DMSO)	7days
Group IV	144.5g	0.0723g of the extract dissolved in 1ml of the carrier (DMSO)	7days
Group V	121.3g	1ml of the carrier (DMSO)	14days
Group VI	106.0g	0.1590g of the extract dissolved in 1ml of the carrier (DMSO)	14days
Group VII	107.6g	0.1076g of the extract dissolved in 1ml of the carrier (DMSO)	14days
Group VIII	95.8g	0.0479g of the extract dissolved in 1ml of the carrier (DMSO)	14days

Myrianthus arboreus (MA) Extracts and the carrier (DMSO) was administered orally using a Gavage syringe (cannula Intubator) on a daily dose.

Throughout the period of study, the animals were observed for any symptom, physiological changes or mortality. Their feed and water were changed daily and good laboratory conditions maintained.

2.2.8. SACRIFICE OF THE ANIMALS

At the end of 7days and 14days respectively, the animals were sacrificed with standard dissection kit. Before sacrificing, the animals were individually sedated (anaesthetized) in a desiccator containing chloroform soaked cotton wool. Blood samples were extracted from the heart and other blood vessels using a syringe. Blood samples were then collected into a well-labelled lithium heparinized blood tubes (to prevent clotting) for Biochemical analysis.

Afterwards, the required organ (liver) was extracted and collected into a sterile (universal) tube containing 10% formalin (to preserve) for histo-pathological studies.

2.2.9 BIOCHEMICAL PARAMETERS STUDIES

The biochemical parameter studies were carried out at Lively Stones Medical Diagnostic Laboratory, Choba, Rivers state.

2.2.9.1 PREPARATION OF BLOOD SAMPLE

Blood samples were put in test tubes of equal volumes and kept exactly opposite each other to balance the weight in a centrifuge. The samples were spurn at 1000r for 5minutes. Thereafter, the supernatant (plasma) was carefully collected with pipette for analysis.

2.2.9.2 LIVER ENZYME ASSAY

The liver enzymes were determined using ELISA Method of determination (Tietz, 1995).

2.2.9.2.1 DETERMINATION OF ALANINE AMINOTRANSFERASE (ALT)

Principle: The principle is based on the reaction of α-ketoglutarate and alanine to yield glutamate and pyruvate in the presence of ALT with dark-brown colour change, which is measured spectrophotometrically.

Sample: Plasma

Reagent composition: R1-Buffer, Phosphate buffer (100mMol/l, pH 7.4), L-alanine (200mMol/l), α-ketoglutarate (2.0mMol/l), R2- 2,4-dinitrophenylhydrazine (2.0mMol/l), R3-sodium hydroxide (0.4mMol/l).

Procedure: Two test tubes were labelled blank and test and 0.1 ml of sample solution and distilled water were added in the two test tubes respectively, 0.5ml of reagent 1 solution (R1) was added to the two test tubes, the mixture were mixed and incubated for 30mins at 37°C in a water bath. After 30mins, 0.5ml of reagent 2 solution (R2) was added to the two test tubes respectively, mixed and further incubated for another 20mins at 25°C, thereafter 5.0ml of reagent 3 solution (R3) was added to both test tubes and mixed. The absorbance of the sample was measured against the reagent black in a spectrophotometer at a wavelength of with 546cm with a 1cm light path cuvette.

Calculation: Absorbance of sample X Standard concentration of ALT (30 u/l)
Absorbance of standard

2.2.9.2.2 DETERMINATION OF ASPARTATE AMINOTRANSFERASE (AST)

Principle: The principle is based on the reaction of diazonium salt with α-oxaloacetate to produce a blue-green complex that is measured spectrophotometrically.

Sample: Plasma

Reagent composition: R1-Buffer, Phosphate buffer (100mMol, pH 7.4), L-aspartate(200mMol/l),α-oxaloacetate(2.0mMol/l),   R2- 2,4-dinitrophenylhydrazine (2.0mMol/l), R3- sodium hydroxide (0.4mMol/l).

Procedure: Two test tubes were labelled blank and test and 0.1 ml of sample solution and distilled water were added in the two test tubes respectively, 0.5ml of reagent 1 solution (R1) was added to the two test tubes, the mixture were mixed and incubated for 30mins at 37°C in a water bath. After 30mins, 0.5ml of reagent 2 solution (R2) was added to the two test tubes respectively, mixed and further incubated for another 20mins at 25°C, thereafter 5.0ml of reagent 3 solution (R3) was added to both test tubes and mixed. The absorbance of the sample was measured against the reagent black in a spectrophotometer at a wave length of with 546cm with a 1cm light path cuvette.

Calculation: Absorbance of sample X Standard concentration of AST (30 u/l)
Absorbance of standard

Principle: The principle is based on the hydrolysis of phenolphthalein monophosposphate, to phosphoric acid and phenolphthalein, which at alkaline pH value turns into a pink colour.

p- Nitrophenylphosphate + H_2O $\xrightarrow{\text{ALP}}$ p-Nitrophenol+ Phosphate.

Sample: Plasma

Reagent composition: R1-Buffer, Diethanolamine buffer (1.0mMol/l; pH9.8), Magnesium (0.6mMol/l), R2- Substrate, p-Nitrophenylphosphate (10mMol/l), R3– Standard.

Procedure: Three test tubes were labelled black, sample and standard, 0.5ml of distilled water was added to black test tube, 1 .0ml of distilled water and 1 drop of agent 1 drop of reagent 1 solution (R1) were also added to both sample and standard test tube respectively, the mixtures were mixed and incubated for 5minutes at 3 7°C. Thereafter, 0.1ml of reagent 3 solution (R3) was added to the standard test tube and 0. 1ml of sample solution was added to sample test tube, thee mixtures were mixed and incubated for another 20minutes at 37°C. After the incubation, 5.0ml of reagent 2 solution (R2) was added to sample and standard test tubes respectively. The mixtures were mixed and the absorbance of the sample was measured against the reagent black in a spectrophotometer at a wavelength of 550nm with a 1cm light path cuvette.

Calculation: $\dfrac{\text{Absorbance of sample}}{\text{Absorbance of standard}}$ X Standard concentration of ALP (120 u/l)

2.2.10 HISTOPATHOLOGICAL STUDIES

The methods involved in histopathological analysis were;

The Fixation Method (Immersion): Before the fixation was done, the volume of fixative was made sufficient to be ten times greater than the volume of tissue to be fixed. 0.9% of saline solution was used to preserved the organ before fixing, the tissues were chopped into smaller pieces as required to aid penetration of the fixative, and drop each pieces into the waiting fixative solution. A serial number was place on top of a strip of white card and dropped into the tube with the tissue as an identifying label. The tubes were corked and inverted a few times to circulate the solution.

Tissue Processing: The tissues were further dehydrated with ethyl alcohol diluted with distil water as required to make a graded series of ethanols, e.g. 50 %, 70 %, and 90% aqueous solution, the dyhydrated tissues were immersed in a suitable organic solvent miscible with both ethanol and parrafin before impregnation can occur successfully. Xylene was further added a clearing agent to lessen the disruptive diffusion currents which usually employs two steps. First step using a mixture of dehydrating and clearing agent and the second step consist purely of clearing agent. The tissues were then transferred to a molten wax, and the wax was changed once or twice to ensure that the clearing agent has been thoroughly replaced by wax in the tissue, and removed from the molten wax surrounding the tissues.

Embedding: This was done by casting the tissues in paraffin wax, allowing them to set to facilitate cutting of sections. Small perforated metal buttons containing the tissues were picked out of the beaker one at a time, the buttons were opened and the identifying labels were noted and kept. A suitable mount was chosen and filled with molten wax with a pair of blunt forceps that has been slightly heated over a Bunsen flame; the tissues were placed in the mould of molten wax with the surface to be cut facing downwards. This forms a tissue block which was thereafter cast onto the wooden block. The specimens were now ready for microtomy.

Microtomy or Sectioning: Microtomes was used to produce very thin sections of specimen that were placed on a microscope slide ready for staining, slides were the prepared and placed in racks. They were warmed daily to avoid lifting of specimen cold.

Staining: Paraffin sections, after they have been cut, were attached onto the slide and stained. Removal of the paraffin wax was achieved with xylene, which is not miscible with watery solution and low-grade alcohol, as such, following treatments of the section with alcohol, the sections were immersed for a few minutes each in 90 % and 70% alcohol, and the sections were then rinsed thoroughly in distilled water. Artefact pigment, present following fixation was removed by the standard method of saturated picric acid. After staining, paraffin wax were mounted in a medium that is miscible with xylene, before passing the sections into two changes of xylene, transparent mounting medium desired to protect the stained section from physical damages was used. A cover slip of appropriate size was cleaned and place on a blotting paper (Whatman N0 1 filter paper), with a blunt forceps, the slides carrying the sections were picked out from the xylene bath, the edges were cleaned or proper visibility of the diamond inscribed number (with a diamond pencil), with a small rod, the necessary amount of mounting medium was placed on the section, slides were inverted and lowered onto the cover slip with gentle pressure, the slides were turned over with a mounted needle, cover slip was squared. Finally, the slides were labelled indicating the tissue stain used and specimen number; the sections were ready to be examined. Hardened mounting medium was achieved by placing slides on hot plates at 50°c.

2.2.11 STATISTICAL ANALYSIS

All data were expressed in mean ± standard error of mean (M ± SEM). Statistical analysis was done by using SPSS software for windows 8™. Data were analysed using analysis of variance (ANOVA) and significant difference of means was determined using Post Hoc and Turkey test for multiple comparisons at the level of $P \leq 0.05$.

CHAPTER THREE

3.1 RESULTS

Table 3.1 Effects of ethanolic leaf extract of *Myrianthus arboreus* (MA) on liver enzymes of Wistar rats for 7days.

GROUPS	EXTRACT ADMINISTERED	ALKALINE PHOPHATASE (ALP) (U/L)	ALANINE TRANSFERASE (ALT) (U/L)	ASPARTEME AMINOTRANSFER ASE (AST) (U/L)
1	DMSO (CONTROL) 7DAYS	21.67 ± .33	18.00 ± 3.00	50.67 ± 5.24
2	1500mg/kg B.W 7DAYS	56.67 ± .88[s]	16.67 ± 2.60	19.67 ±4.06 [s]
3	1000mg/kg B.W 7DAYS	36.00 ± 1.73 [s]	10.67 ± 1.33	47.33 ± 6.64
4	500mg/kg B.W 7DAYS	30.67 ± 1.45 [s]	12.33 ± 2.60 [s]	41.33 ± 6.07

Results represented in Mean ± Standard error of mean (Mean ± SEM)

In Table 3.1 superscript "s" represents a significant difference between groups when compared with the DMSO control group.

Table 3.1 shows that in groups 2, 3 & 4, the *ALP* values has Significant increase ($P \leq 0.05$) when compared with the DMSO control group (i.e. group 1).

Table 3.1 shows that the *ALT* values of groups 2, 3 & 4 have a non-significant difference ($P \geq 0.05$) when compared with the DMSO control group (i.e. group 1).

Table 3.1 shows that the *AST* values of groups 3 & 4 have a non-significant difference ($P \geq 0.05$) in *AST* values but only group 2 has Significant reduction ($P \leq 0.05$) when compared with the DMSO control group (i.e. group 1).

Table 3.2 Effects of ethanolic leaf extract of *Myrianthus arboreus* (MA) on liver enzymes of wistar rats for 14days.

GROUPS	EXTRACT ADMINISTERED	ALKALINE PHOPHATASE (ALP) (U/L)	ALANINE TRANSFERASE (ALT) (U/L)	ASPARTEME AMINOTRANSFER SE (AST) (U/L)
5	DMSO (CONTROL) 14DAYS	40.33± .67	8.00 ± .00	36.00 ± 2.89
6	1500mg/kg B.W 14DAYS	52.67 ± .67[s]	35.67 ± 1.67 [s]	29.67 ± 1.33
7	1000mg/kg B.W 14DAYS	47.33 ± 2.60	26.33 ± 1.33 [s]	48.33 ± 10.41
8	500mg/kg B.W 14DAYS	91.33 ± 2.73 [s]	26.33 ± 1.33 [s]	52.67 ± 3.48

Results represented in Mean ± Standard error of mean (Mean ± SEM)

In Table 3.2, superscript "s" represents a significant difference between groups when compared with the DMSO control group.

Table 3.2 shows groups 6 & 8 has Significant increase at ($P \leq 0.05$) when compared with the DMSO control group (i.e. group 5). Only group 7 has a Non-Significant difference ($P \geq 0.05$) in *ALP* values when compared with the DMSO control group (i.e. group 5).

Table 3.2 shows significant increase at ($P \leq 0.05$) exist in *ALT* values of Groups 6, 7 & 8, when compared with the DMSO control group (i.e. group 5).

Table 3.2 shows groups 6, 7 & 8, have a non-significant difference ($P \geq 0.05$) in *AST* values when compared with the DMSO control group (i.e. group 5).

3.2 RESULTS OF HISTOLOGICAL ANALYSIS

The plates below shows the histology of the liver of wistar rat after administering DMSO and ethanolic leaf extracts of *Myrianthus arboreus* (MA) for 7 and 14 days.

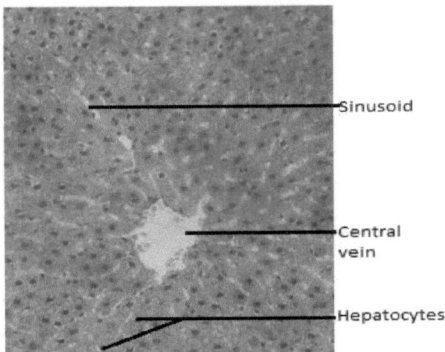

PLATE 3.1. HISTOLOGY OF THE LIVER TISSUE OF WISTAR RAT ADMINISTERED WITH DMSO ONLY FOR 7DAYS (H&E STAINING) x400

INTERPRETATION OF RESULT: The result shows that the sinusoid (a small blood vessel or cavity in the tissue of an organ such as the liver), the central vein and the hepatocytes are all normal.

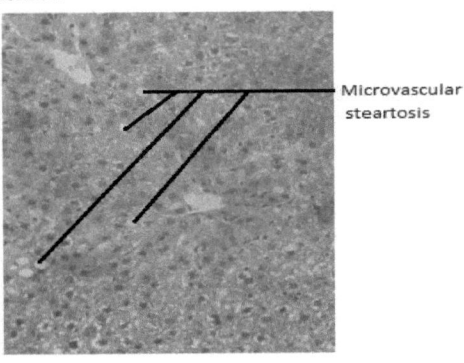

PLATE   3.2. HISTOLOGY OF THE LIVER TISSUE OF WISTAR RAT ADMINISTERED WITH DMSO + 1500mg/Kg B.W OF ETHANOLIC LEAF EXTRACT OF *MYRIANTHUS ARBOREUS* FOR 7DAYS (H&E STAINING) x400

INTERPRETATION OF RESULT: The result shows the occurrence of Microvesicular Steatosis (also called microvesicular fatty degeneration or microvesicular adipose degeneration). It is the process describing the abnormal retention of lipids within a cell.

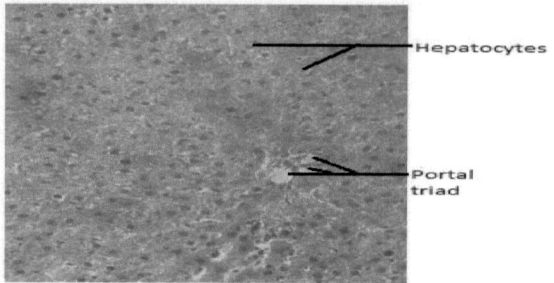

PLATE   3.3. HISTOLOGY OF THE LIVER TISSUE OF WISTAR RAT ADMINISTERED WITH DMSO + 1000mg/Kg B.W OF ETHANOLIC LEAF EXTRACT OF *MYRIANTHUS ARBOREUS* FOR 7DAYS (H&E STAINING) x400

INTERPRETATION OF RESULT: The result shows a normal liver. Portal triad is a distinctive arrangement in the liver.

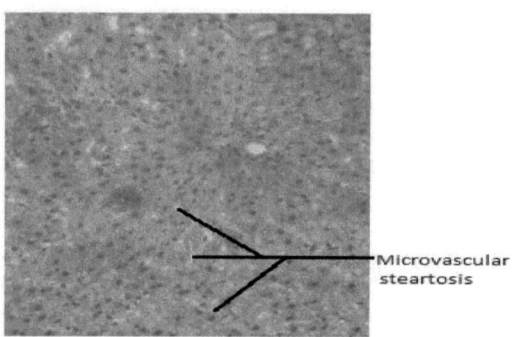

PLATE   3.4. HISTOLOGY OF THE LIVER TISSUE OF WISTAR RAT ADMINISTERED WITH DMSO + 500mg/Kg B.W OF ETHANOLIC LEAF EXTRACT OF *MYRIANTHUS ARBOREUS* FOR 7DAYS (H&E STAINING) x400

INTERPRETATION OF RESULT: The result shows the occurrence of Microvesicular Steatosis (also called microvesicular fatty degeneration or microvesicular adipose degeneration).

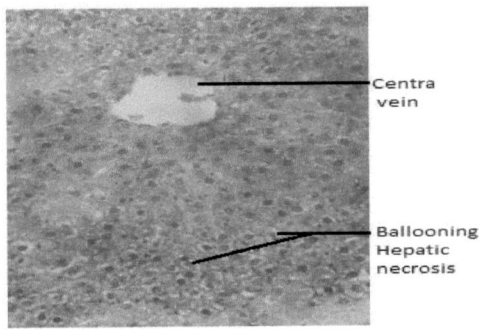

PLATE   3.5. HISTOLOGY OF THE LIVER TISSUE OF WISTAR RAT ADMINISTERED WITH DMSO + 1000mg/Kg B.W OF ETHANOLIC LEAF EXTRACT OF *MYRIANTHUS ARBOREUS* FOR 14DAYS (H&E STAINING) x400

INTERPRETATION OF RESULT: The result shows a ballooning degeneration of the liver parenchymal cell (i.e. hepatocytes).

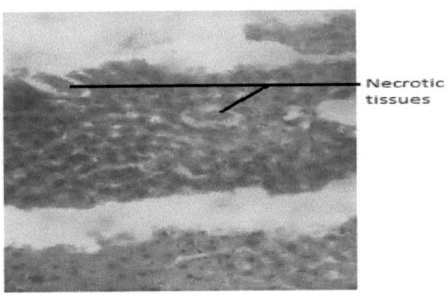

PLATE   3.6. HISTOLOGY OF THE LIVER TISSUE OF WISTAR RAT ADMINISTERED WITH DMSO + 500mg/Kg B.W OF ETHANOLIC LEAF EXTRACT OF *MYRIANTHUS ARBOREUS* FOR 14DAYS (H&E STAINING) x400

INTERPRETATION OF RESULT: The result shows lesions within a circumscribed area of the tissue which explains necrosis of the tissue.

3.3 RESULTS OF PHYTOCHEMICAL SCREENING OF THE CRUDE LEAF POWDER OF *MYRIANTHUS ARBOREUS* (P.BEAUV)

Table 3.3. Results of Phytochemical Screening

PHYTOCHEMICAL	TEST	RESULTS
Alkaloids	Drangedorff's Test	-ve
	Mayer's Test	+ve
	Hager's Test	+ve
Flavonoids	Shinodu's Test	+ve
	Leadacetate Test	+ve
	NaOH Test	+ve
Tannins	FeCl$_3$ Test	+ve
	Phlobatannins	-ve
Anthraquinone	Free Anthroquinone	-ve
	Combined Anthroquinone	-ve
Triterpenoids	Liebermann-Buchard Test	-ve
	Salwoski Test	+ve
Carbonhydrates	Molisch's Test	+ve
	Fehling's Test	+ve
Cardenolide	Keller-killani Test	-ve
	Kedde Test	-ve
Saponins	Frothing Test	+ve
	Emulsion	+ve
	Na$_2$CO$_3$	-ve
Cyanogenic glycosides	___	ND
Fixed oils	___	ND

ND means metabolites Not Determined

-ve means absent

+ve means present

CHAPTER FOUR

4.0   DISCUSSIONS AND CONCLUSIONS

4.1   DISCUSSIONS

In the review, numerous nutritional and health benefits of *Myrianthus arboreus* was pointed out. Moreover, owing to the fact that there was no available toxicological data regarding the safety of repeated exposure to this plant, this study was necessitated.

With respect to the phytochemical screening of the crude leaf of *Myrianthus arboreus* (Table 3.3), several secondary metabolites were identified (which includes alkaloids, flavonoids, tannins, anthroquinone, triterpenoids, carbonhydrates, cardenolide and saponins). These bioactives comounds upon extraction from the plant material and administration to experimental animals is seen as a Xenobiotic and exposes vital organs of the body such as Liver, kidney, Heart, spleen, Lungs and Pancreas to damage.(Dybing *et al.*, 2002).

The liver is known to be the main site of detoxification of xenobiotic as such, serves as filtration of toxins. Hence, it is more prone to damage. (Wikipedia, 2015).

The ethanolic leaf extract of *Myrianthus arboreus* was administered to the rat at Lethal Doses (LD$_{50}$), observed for 7days and no death was recorded. Therefore, a single oral dose of 5000mg/Kg Body weight, 3000mg/kg Body weight and 1000mg/kg body weight of ethanolic leaf extract of *Myrianthus arboreus* did not result to any toxic effects in the Lethal doses.

The results of Biochemical analysis (Tables 3.1 and 3.2) shows an elevation in plasma enzyme levels such as Alkaline phosphatase (ALP), Alanine Transaminase (ALT) and Aspartate Aminotransferase (AST) when compared with the effects of the DMSO control (the carrier) for 7 and 14days.

AST is found in decreasing concentration in the liver, cardiac muscle, Skeletal muscle, kidney, brain, pancreas, lungs, leukocytes and erythrocytes; whereas ALT is primarily found in the liver and kidney. Thus, an elevation of ALT is more specific for liver injury than elevation of AST. (ASCP, 2003).
Apart from that, AST values actually showed a non-significant difference ($P \geq 0.05$) when compared with the DMSO control groups.
The results of histological analysis (Plates 3.2 and 3.4) show microvesicular steatosis of the liver (occurring at a higher dosage of administering the extract), which is not particularly detrimental to the cell especially in mild cases. (Wikipedia, 2015).
The results (Plate 3.5) also show ballooning hepatic necrosis and necrotic tissues (Plate 3.6), occurring at higher dosage and longer time of administering the extract. The reason for this histological status could be owing to the presence of phytochemicals in plant such as Alkaloids, Phenolic and tannins(Encyclopeadia britanica, 2013).
Daily oral administration of the extract for 14days (acute toxicity studies) did not also cause mortality.
Hence, the No-observed adverse-effect level of the extract was found to be below 500 mg/kg/day.

4.2   CONCLUSIONS

In this study, we can conclude that the ethanolic leaf extract of *Myrianthus arboreus* was unsafe at all doses considered for a period of 14days. However, at a dose below 500mg for 7days could be considered safe.

4.3   LIMITATIONS TO THE STUDY

Limitations experienced during the study were insufficient space for the animals, poor laboratory conditions of the animal house and unavailability of standard laboratory equipment for the research.

4.4   SUGGESTIONS FOR FURTHER RESEARCH

A sub-acute and chronic toxicity tests of the extract of *Myrianthus arboreus* plant is suggested.

In addition, the test should be carried out by isolating a specific phytochemical of the plant to administer to the animals.

REFERENCE

Aashish, P., Tarun, S. and Pallavi, B. (2011). Drug Induced Hepatotoxicity: A Review. *Journal of Applied Pharmaceutical Science* 02 (05); 233-243

Agwa, O.K., Chuku, W. and Obichi, E.A (2011).The in vitro effect of *Myrianthus arboreus* leaf extract on some pathogenic bacteria of clinical origin. *Journal of Microbiology and Biotechnology Research,2011, 1 (4):77-85.*

Agyare, C., Ansah, A.O., Ossei, P.P.S., Apenteng J.A. and Boakye, Y.D. (2014).Wound Healing and Anti-Infective Properties of *Myrianthus arboreus* and Alchornea cordifolia. Med chem 4: 533-539. doi:10.4172/2161-0444.1000191.

Andy, S. (2001). Hepatic Toxicity. MSc in Molecular Pathology and Toxicology p.3.

Ayoola, A.A., Efeovbokhan, V.C., Bafuwa, O.T.and David, O.T. (2014). A Search for Alternative Solvent to Hexane During Neem Oil Extraction. *International Journal of Science and Technology Volume 4 No. 4, April 2014* p. 69.

ASCP 2003 educational commentary.

Benjamin, L.S., and Sherman, P.M (2008). Peadiatric coustromtestmal disease. Connection: PMPH-USA p.751.

Bhardwaj, S., Deepika, G., Seth, G. L. and Bihani, S.D. (2012). Study Of Acute, Sub Acute and Chronic Toxicity Test. *International Journal of Advanced Research in Pharmaceutical and Biosciences,* V.2 (2):103-129.

Crook, M.A. (2006). Chemical chemistry and metabolic medicine (7th ed.). Bokk power publishers, Indian P232 contra, kumat, V. Fauto, N., Robbins, S.L.E.

Chin, Y.W, Balunas, M.J, Cha,I. H.B. and Kinghorn, A.D.(2006).Drug discovery from Natural Sources. *The AAPS Journal.*; 8(2):E239-53.

De Pasquale, A. (1984). Pharmacognosy:The Oldest Modern Science. *Journal of Ethnopharmacology, 11, 1–16.*

Dybing, E., Doe, J., Groten, J., Kleiner, J., and O'Brien, J. (2002). Hazard characterization of chemicals in food and diet: dose response, mechanism and extrapolation issues. *Food Chem. Toxicol.* 2002, 42, 237-282.

Encyclopædia Britannica (2013). Necrosis. *Encyclopædia Britannica Ultimate Reference Suite.* Chicago: Encyclopædia Britannica.

Encyclopædia Britannica (2013). Alkaloid. *Encyclopædia Britannica Ultimate Reference Suite.* Chicago: Encyclopædia Britannica.

Enzyme markers of Enzyme markers of Toxicity' Retrieved from www.google.com on 2015-09-04.

Iliuţã, A. (2011). Experimental Use of Animals in Research. *Balneo-Research Journal, V.2, Pp.1-5.*

IUPAC, *Compendium of Chemical Terminology*, 2nd ed. (the "Gold Book") (1997). Online corrected version: (2006–) "acute toxicity".

Kmiec, Z (2001). "Cooperation of liver cells in health and disease". Advance Anatomy of Embryo cell biology, 161: III-XIII, 1-151.

Koolman, J. and Roehm, k.H. (2005). Cell fragmentation: Marker molecules. *Color Atlas of Biochemistry,* 2nd edition p.198.

Leung, W.T.W., Busson, F. and Jardin, C. (1968). Food Composition Table for Use in Africa. FAO, Rome, Italy. p.306.

Lynch, T., E. Prince, A. (2007), The effect of cytochrome P450 metabolism on drug response, "interaction and adverse effects" *American family physicians,*76 (3): 391-6.

Microsoft Encarta,2009.

Mouhssen, L.(2013). The Success of Natural Products in Drug Discovery *Pharmacology and Pharmacy*, 2013, 4, pp. 17-31 http://dx.doi.org/10.4236/pp.2013.43A003 Published Online June 2013 (http://www.scirp.org/journal/pp).

Newman, D.J. and Cragg, G.M. (2012). Natural Products as Sources of New Drugs over the 30 Years from 1981 to 2010. *Journal of Natural Products*, 75(3):311-35.

OECD Guidelines for Testing Chemicals (2008).

Okafor, J.C. (2004). *Myrianthus arboreus* P.Beauv. In: Grubben, G.J.H. and Denton, O.A. (Editors). *PROTA 2: Vegetables/Légumes*. [CD-Rom]. PROTA, Wageningen, Netherlands.

Oyeyemi, S. D., Arowosegbe, S. and Adebiyi, A. O. (2014). Phytochemical and Proximate Evaluation of *Myrianthus arboreus* (P.Beau.) and Spargonophorus Sporgonophora (Linn.) Leaves. *Journal of Agriculture and Veterinary Science Volume 7, Issue 9 Version I (Sep. 2014), PP 01-05.*

Pak, E., Esrason, K.T., and EWU, V.H. (2004). "Hepatotoxicity of herbal remedies": "an emerging dilemma". *Progress in Transplantation*, 14(2): 91-6.

Prasanth, K., Suba, V., Ramireddy, B., and Srinivas B. (2014). Acute and Sub-Acute (28-Day) Oral Toxicity Studies of Ethanolic Extract of *Celtis Timorensis* Leaves in Rodents: *Global Journal of Medical Research V 14:3* P.7.

Raza, M., Al-Shabanah, O.A., El-Hadiyah, T.M. and Al-Majed, A.A. (2002). Effect of prolonged vigabatrin treatment on hematological and biochemical parameters in plasma, liver and kidney of Swiss albino mice. Sci. Pharm. 2002, 70, 135-145.

Saad, B., Azaizeh, H., Abu-Hijleh, G. and Said, O. (2006). Safety of Traditional Arab Herbal Medicine. *Evidence Based Complement Alternate Medicine*. 2006, 3:433-9.

Segura, G. (2011). DMSO - The Real Miracle Solution Retrieved on 16 October 2015 from www.sott.net.

Sharma, A., Chakraborti, K.K. and Handa, S.S. (1991). Anti-hepatotoxic Activity Of Some Indian Herbal Formulations as Compared To Silymarin. Fitoterapia. 1991; 62: 229-235.

The MSDS HyperGlossary: Acute toxicity. Safety Emporium. Archived from the original on 16 October 2006. Retrieved 2006-11-15.

Tietz, N.W. (1995). Clinical guide to laboratory tests. Philadephia: W.B Saunders company.

Walum, E. (1998). "Acute oral toxicity". *Environmental Health Perspectives* Vol. 106 (2): 497–503.

Wikipedia, The free encyclopaedia (2015) Retrieved on 2015-09-06.

Www.Wikipeadia.Com. (2015).Detoxification. Retrieved on 2 December 2015.

APPENDIX 1

PREPARATION OF THE ETHANOLIC EXTRACT STUCK
PREPARATION OF EXTRACT STUCK FOR LD$_{50}$:

Weights of Animals for LD$_{50}$

Group A: 263.2g, 259.4g & 260.3g   (Average weight = 261.0g)

Group B: 236.6g, 234.4g & 237.0g   (Average weight = 236.0g)

Group C: 164.2g, 163.2g & 167.1g   (Average weight = 164.8g)

Calculations for LD$_{50}$ Administration

Group A: (5000mg/Kg Body weight) (3rats)

5000mg (of extract) $\longrightarrow$ 1kg = 1000g (rat)

Xmg (of extract) $\longrightarrow$ 261.0g (rat)

Where Xmg = quantity of required

Therefore, Xmg (of extract) $= \dfrac{5000\text{mg (of extract) x }261.0\text{g (rat)}}{1000\text{g (rat)}} = 1305\text{mg} = 1.305\text{g}$

Hence, 1.305g of the extract was dissolved in 1ml of the carrier (DMSO)

Group B: (3000mg/Kg Body weight) (3rats)

3000mg (of extract) $\longrightarrow$ 1kg = 1000g (rat)

Xmg (of extract) $\longrightarrow$ 236.0g (rat)

Therefore, Xmg (of extract) $= \dfrac{3000\text{mg (of extract) x }236.0\text{g (rat)}}{1000\text{g (rat)}} = 708\text{mg} = 0.708\text{g}$

Hence, 0.708g of the extract was dissolved in 1ml of the carrier (DMSO)

Group C: (1000mg/Kg Body weight) (3rats)

1000mg (of extract) $\longrightarrow$ 1kg = 1000g (rat)

Xmg (of extract) $\longrightarrow$ 164.0g (rat)

Therefore, Xmg (of extract) $=$ $\dfrac{1000\text{mg (of extract) x } 164.0\text{g (rat)}}{1000\text{g (rat)}}$ $=$ 164mg = 0.164g

Hence, 0.164g of the extract was dissolved in 1ml of the carrier (DMSO)

PREPARATION OF EXTRACT STUCK FOR ACUTE TOXICITY STUDIES:

Weights of Animals for acute toxicity studies

Group 1: Average weight = 121.3g.

Group 2: Average weight = 123.3g.
Group 3: Average weight = 120.5g

Group 4: Average weight = 144.5g

Group 5: Average weight = 121.3g
Group 6: Average weight = 106.0g

Group 7: Average weight = 107.6g

Group 8: Average weight = 95.8g

Calculations for Acute toxicity Administration

Group 1: Administered with 1ml of concentrated DMSO for 7days

Group 2: (1500mg/Kg Body weight) (3rats)

15000mg (of extract) $\longrightarrow$ 1kg = 1000g (rat)

Xmg (of extract) $\longrightarrow$ 123.3g (rat)

Where Xmg = quantity of required

Therefore, Xmg (of extract) $=$ $\dfrac{1500\text{mg (of extract) x } 123.3\text{g (rat)}}{1000\text{g (rat)}}$ = 184.95mg = 0.18495g

Hence, 0.18495g of the extract was dissolved in 1ml of the carrier (DMSO) and administered orally for 7days.

Group 3: (1000mg/Kg Body weight) (3rats)

1000mg (of extract) $\longrightarrow$ 1kg = 1000g (rat)

Xmg (of extract) $\longrightarrow$ 120.5g (rat)

Where Xmg = quantity of required

Therefore, Xmg (of extract) $=$ $\dfrac{1000\text{mg (of extract) x } 120.5\text{g (rat)}}{1000\text{g (rat)}}$ = 120.5mg = 0.1205g

Hence, 0.1205g of the extract was dissolved in 1ml of the carrier (DMSO) and administered orally for 7days.

Group 4: (500mg/Kg Body weight) (3rats)

500mg (of extract)   ——————————————→   1kg = 1000g (rat)

      Xmg (of extract)   ——————————————→   144.5g (rat)

Where Xmg = quantity of required

Therefore, Xmg (of extract) = $\dfrac{\text{500mg (of extract) x 144.5g (rat)}}{\text{1000g (rat)}}$ = 72.25mg = 0.07225g

Hence, = 0.07225g of the extract was dissolved in 1ml of the carrier (DMSO) and administered orally for 7days.

Group 5: Administered with 1ml of concentrated DMSO for 14days

Group 6: (1500mg/Kg Body weight) (3rats)

1500mg (of extract)   ——————————————→   1kg = 1000g (rat)

      Xmg (of extract)   ——————————————→   106.5g (rat)

Where Xmg = quantity of required

Therefore, Xmg (of extract) = $\dfrac{\text{1500mg (of extract) x 106.5g (rat)}}{\text{1000g (rat)}}$ = 159.75mg = 0.15975g

Hence, 0.15975g of the extract was dissolved in 1ml of the carrier (DMSO) and administered orally for 14days.

Group 7: (1000mg/Kg Body weight) (3rats)

1000mg (of extract)   ——————————————→   1kg = 1000g (rat)

      Xmg (of extract)   ——————————————→   107.6g (rat)

Where Xmg = quantity of required

Therefore, Xmg (of extract) = $\dfrac{\text{1000mg (of extract) x 107.6g (rat)}}{\text{1000g (rat)}}$ = 107.6mg = 0.1076g

Hence, 0.1076g of the extract was dissolved in 1ml of the carrier (DMSO) and administered orally for 14days.

Group 7: (1000mg/Kg Body weight) (3rats)

1000mg (of extract) $\longrightarrow$ 1kg = 1000g (rat)

    Xmg (of extract) $\longrightarrow$ 107.6g (rat)

Where Xmg = quantity of required

Therefore, Xmg (of extract) $=$ $\dfrac{1000\text{mg (of extract)} \times 107.6\text{g (rat)}}{1000\text{g (rat)}} = 107.6\text{mg} = 0.1076\text{g}$

Hence, 0.1076g of the extract was dissolved in 1ml of the carrier (DMSO) and administered orally for 14days.

Group 8: (500mg/Kg Body weight) (3rats)

500mg (of extract) $\longrightarrow$ 1kg = 1000g (rat)

    Xmg (of extract) $\longrightarrow$ 95.8g (rat)

Where Xmg = quantity of required

Therefore, Xmg (of extract) $=$ $\dfrac{500\text{mg (of extract)} \times 95.8\text{g (rat)}}{1000\text{g (rat)}}$ $=$ 47.9mg $=$ 0.0479g

Hence, 0.0479g of the extract was dissolved in 1ml of the carrier (DMSO) and administered orally for 14days.

APPENDIX 2

RESEARCH TOPIC:

ACUTE TOXICITY OF ETHANOLIC LEAVE EXTRACT OF MYRIANTHUS ARBOREUS ON WISTAR ALBINO RATS.

RESEARCH OBJECTIVES:
- ❖ **DETERMINATION OF LETHAL DOSE (LD_{50})**
- ❖ **QUALITATIVE PHYTOCHEMICAL SCREENING OF THE PLANT**
- ❖ **ACUTE TOXICITY ON THE LIVER; AST, ALP & AST.**

RESEARCH METHODOLOGY:

LD_{50} DETERMINATION ; (3 GROUPS WITH 3 RATS EACH LABELLED A, B & C)
- ❖ GROUP A ADMINISTERED WITH 5000MG/KG BODY WEIGHT.
- ❖ GROUP B ADMINISTERED WITH 3000MG/KG BODY WEIGHT.
- ❖ GROUP C ADMINISTERED WITH 1000MG/KG BODY WEIGHT.

ACUTE TOXICITY STUDIES; (8 GROUPS WITH 3 RATS EACH LABELLED I- VIII)
- ❖ GROUP I   CONTROL; ADMINISTERED CARRIER FOR 7DAYS.
- ❖ GROUP II   ETHANOLIC EXTRACT OF x DOSE FOR 7DAYS.
- ❖ GROUP III   ETHANOLIC EXTRACT OF y DOSE FOR 7DAYS.
- ❖ GROUP IV   ETHANOLIC EXTRACT OF z DOSE FOR 7DAYS.
- ❖ GROUP V   CONTROL; ADMINISTERED CARRIER FOR 14DAYS.
- ❖ GROUP VI   ETHANOLIC EXTRACT OF x DOSE FOR 14DAYS.
- ❖ GROUP VII   ETHANOLIC EXTRACT OF y DOSE FOR 14DAYS.
- ❖ GROUP VIII   ETHANOLIC EXTRACT OF z DOSE FOR 14DAYS

YOUR KNOWLEDGE HAS VALUE

- We will publish your bachelor's and master's thesis, essays and papers

- Your own eBook and book - sold worldwide in all relevant shops

- Earn money with each sale

Upload your text at www.GRIN.com
and publish for free

YOUR KNOWLEDGE HAS VALUE